Exploring Shapes™

Pyramids

Bonnie Coulter Leech

The Rosen Publishing Group's
PowerKids Press™
New York

To Bill and Billy—the wind beneath my wings

Published in 2007 by The Rosen Publishing Group, Inc.
29 East 21st Street, New York, NY 10010

First Edition

Editor: Kara Murray
Book Design: Elana Davidian
Layout Design: Greg Tucker

Illustrations: pp. 6, 8, 9, 10, 11, 12, 14, 16, 19 by Greg Tucker.
Photo Credits: Cover © Roger Ressmeyer/Corbis; p. 5 © Yann Arthus-Bertrand/Corbis; p. 7 © SIME/SIME/Corbis; p.13 © Mike E. Gibson/Corbis; p. 15 © Jeremy Horner/Corbis; p. 17 © Gail Mooney/Masterfile; p. 21 © Steve Vidler/Superstock.

Library of Congress Cataloging-in-Publication Data

Leech, Bonnie Coulter.
  Pyramids / Bonnie Coulter Leech.— 1st ed.
     p. cm. — (Exploring shapes)
  Includes index.
  ISBN 1-4042-3499-3 (lib. bdg.)
  1. Polygons—Juvenile literature. 2. Geometry, Plane—Juvenile literature. 3. Pyramids—Juvenile literature. I. Title. II. Series.
  QA482.L43794 2007
  516'.156—dc22
                                          2006000290

Manufactured in the United States of America

Contents

Pyramids of Ancient Egypt

If you say the word "pyramid," most people will think of ancient Egypt. The ancient Egyptians built many famous pyramids. The Great Pyramid of Giza, the largest of the Egyptian pyramids, was built nearly 4,500 years ago and stands 481 feet (147 m) high. It is located in northern Egypt near the city of Cairo.

The Great Pyramid of Giza was built to be the grave for King Khufu. Pyramids provided a place where a king's body could safely pass into the **afterlife**. Many great riches were held in rooms within the pyramids.

In mathematics we study **three-dimensional** shapes that look like the ancient pyramids. These geometric shapes are also called pyramids.

The Great Pyramid is the largest of the three pyramids of Giza, in northern Egypt. It is also known as Kheops or Khufu, for the king who is buried there. It is thought that it took about 20 years to build the Great Pyramid.

We live in a three-dimensional world. Many things around us have length, width, and height. Any shape that has length, width, and height is called a solid figure. There are many different types of solid shapes and figures around us.

Solid figures do not lie in a plane. A plane is a flat surface. Shapes that lie in a plane are two-dimensional, which means that they have only two measurements, or dimensions. Two-dimensional shapes have length and width.

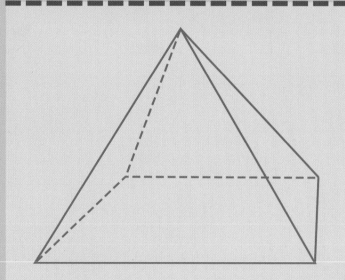

Pyramids are solid figures. They have length, width, and height.

Look again at the pyramid on page 6. Notice how there is a shape upon which the pyramid sits. That shape is the base. In the pyramid on page 6, the base is a rectangle. The sides of the pyramid are triangles. The pyramids we study in mathematics have one base. The base can be any **polygon**, and the sides are always triangles.

This pyramid stands in Rome, Italy. The Roman ruler Caius Cestius was buried here in 12 B.C.

A pyramid has only one base. The base of a pyramid can be any kind of polygon. A polygon is a closed, two-dimensional figure. Polygons are closed figures because they have no openings. All their sides meet another side.

A pyramid is named for the polygon that makes up its base. The three-dimensional pyramid will have the same number of sides as the polygon that makes up its base. A triangle is a polygon with

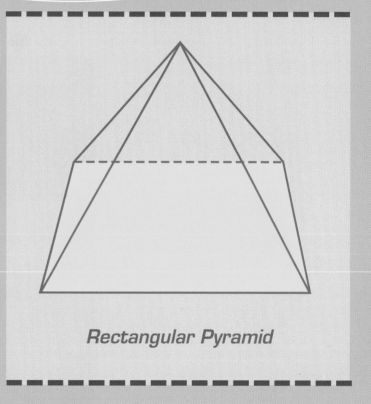

Rectangular Pyramid

three sides. A pyramid with a base that is a triangle is called a triangular pyramid. That pyramid will have three sides.

A pyramid with a square base is called a square pyramid. A square is a polygon with four **congruent** sides and four right angles. A right angle is like the corner of a sheet of paper. The sides of a square pyramid are four congruent triangles.

All regular pyramids have congruent triangles for sides. In this book we will be studying regular pyramids.

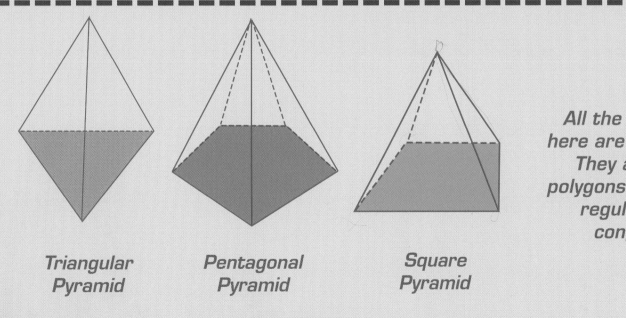

Triangular
Pyramid

Pentagonal
Pyramid

Square
Pyramid

All the pyramids shown here are regular pyramids. They all have regular polygons for their bases. A regular polygon has congruent sides.

The base and the sides of a pyramid are called **faces**. The base can be any polygon, but the sides are all triangles. The triangular faces that make up the sides of a pyramid are called **lateral faces**. The word "lateral" means "side."

The faces of a pyramid meet to form a line segment. A line segment is part of a line with two endpoints. If you look at the place where two walls meet, you will see a line

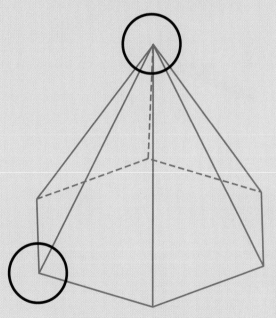

The top of a pyramid is called the apex. It is also a vertex.

The edges of a pyramid meet at a vertex.

segment. In a pyramid the line segments that make up the faces are called **edges**.

The edges of a lateral face are called **lateral edges**. The edge where a triangular lateral face joins the base is called a base edge.

The place in a room where two walls meet the floor is called a corner. In a pyramid the point where three or more edges meet is also called a corner. A corner is also called a **vertex**. Vertices are what we call more than one vertex.

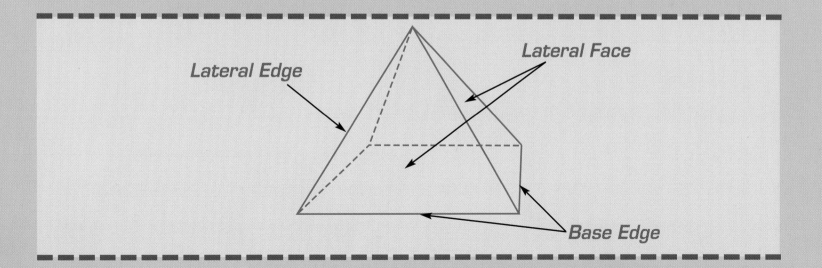

Lateral Edge

Lateral Face

Base Edge

The word "**altitude**" is often used to describe how high an airplane is flying in the sky. Altitude also tells us the height of a mountain. Only three-dimensional figures have height. Two-dimensional figures have only length and width, but they do not have height.

Altitude

This is used to show that the altitude makes a right angle with the base.

Pyramids and other solid figures have height. The height of a pyramid is called the altitude. The altitude of a pyramid is the line segment that goes from the top vertex, or **apex**, to the center of the base of the pyramid. The altitude of a pyramid forms a right angle with the base.

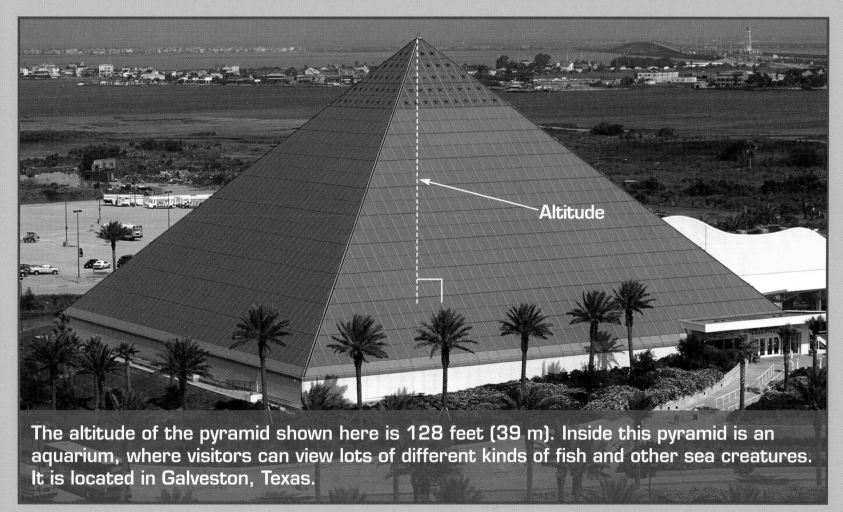

Altitude

The altitude of the pyramid shown here is 128 feet (39 m). Inside this pyramid is an aquarium, where visitors can view lots of different kinds of fish and other sea creatures. It is located in Galveston, Texas.

When we talk about a pyramid, whether it is one of the ancient pyramids of Egypt or one of the other pyramids in this book, we are always talking about regular pyramids. In a regular pyramid the base is a polygon that has all sides congruent. The lateral faces are congruent triangles and the lateral edges are all congruent. In addition, in a regular pyramid the apex is located directly over the center of the base.

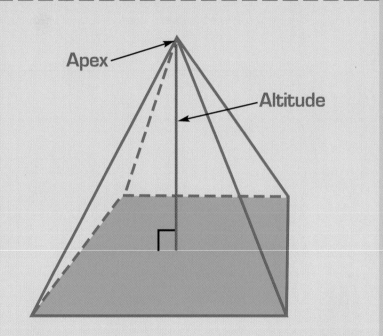

As in all regular pyramids, the altitude of this regular pyramid forms a right angle at the center of the base.

This regular pyramid was formed by cannon balls. You can tell it is a regular pyramid because each side of the base is made up of six cannon balls. Each side is congruent. It is a square pyramid. You could try to make a regular pyramid out of tennis balls or golf balls.

In a pyramid there are two types of altitudes. One altitude tells us the height of the pyramid. This altitude is drawn from the apex to the base. The second type of altitude is the height of each of the triangular lateral faces.

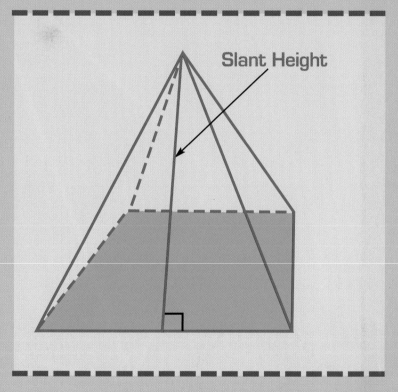

In a triangle the altitude is the line segment drawn from the vertex to the base. The altitude forms a right angle with the base.

Each of the lateral faces of a pyramid is a triangle. Each of these triangular faces has an altitude. The altitude of a lateral face of a pyramid is called the slant height. The lateral faces of the pyramid slant, or lean. The slant height of a lateral face is the line drawn from the top vertex to the base of the triangular lateral face. The slant height forms a right angle with the base of the triangle that forms the lateral face.

Slant Height

This is the Pyramid Arena in Memphis, Tennessee. Each of its lateral faces has a slant height of 541 feet (165 m). The Pyramid Arena is the third-largest pyramid in the world.

If you take a shoe box and cut along its sides so that it lays flat on the surface of a table, you can see each of the shapes that make up the box. All the different parts that make up a box are rectangles. When you cut along the edges of a solid figure and flatten it to show the shapes of which it is made, you are looking at a **net**. A net is a closed plane figure that can be folded into a three-dimensional shape.

Nets are patterns, or two-dimensional pictures, of a three-dimensional shape. These two-dimensional patterns can be folded to make a three-dimensional solid. Pyramids have nets. The shape of the net depends on how the pyramid is unfolded. Think of all the different ways you can unfold a shoe box. You could unfold a pyramid just as many ways.

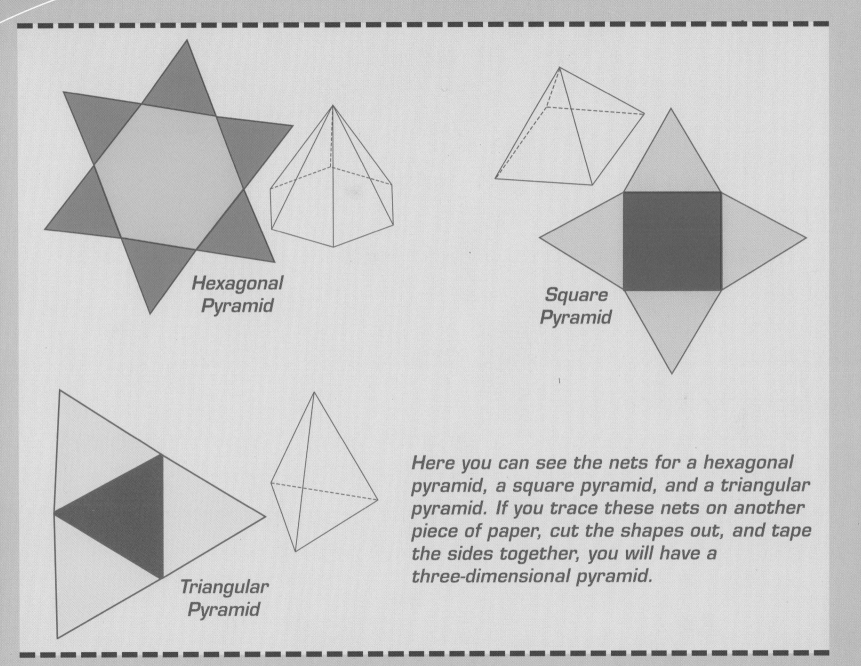

Hexagonal Pyramid

Square Pyramid

Triangular Pyramid

Here you can see the nets for a hexagonal pyramid, a square pyramid, and a triangular pyramid. If you trace these nets on another piece of paper, cut the shapes out, and tape the sides together, you will have a three-dimensional pyramid.

A net is helpful in finding the **surface area** of a solid. The surface area is the sum of the areas of all the faces of a solid. In a pyramid the surface area is the area of the base plus the area of all the lateral faces. If you were going to paint the surface area of a pyramid, you would paint each of the triangular faces and the base. You would then have painted the total surface area of the pyramid.

What if you just wanted to paint the sides of a pyramid, but not the base? If you painted just the surface of the lateral faces, you would have painted the lateral surface area. The area of the lateral faces of a pyramid is the lateral surface area.

The pyramid shown here is at the front of the Louvre Museum in Paris, France. If you count all the diamonds on the outside of the pyramid, you will find the lateral surface area measured in diamonds. The shapes on the bottom row are triangles, which are half diamonds. There are 603 diamonds and 70 triangles. Therefore, the lateral surface area is 638 diamonds.

Pyramids Around Us

Many great inventors and **architects** have used pyramids in their **designs**. Architects are people who design buildings. I. M. Pei is one architect who uses pyramids in his work. Pei designed the glass and steel pyramid for the entrance to the Louvre Museum in Paris, France. Alexander Graham Bell used pyramids for his designs of kites. Around 1485, Leonardo da Vinci designed a parachute in the shape of a pyramid.

I. M. Pei, Alexander Graham Bell, Leonardo da Vinci, and the ancient Egyptians recognized the beauty and strength of the pyramid. For thousands of years, the pyramids in Egypt have stood for the accomplishments of the ancient Egyptians. On the back of a U.S. one-dollar bill, there is a picture of a pyramid. This pyramid stands for the strength and stability of the United States.

Glossary

afterlife (AF-ter-lyf) Where people believe they will go after they die.

altitude (AL-tih-tood) The height above Earth's surface.

apex (A-peks) The top or uppermost part.

architects (AR-kih-tekts) People who create ideas and plans for a building.

congruent (kun-GROO-ent) Having the same measurement and shape.

designs (dih-ZYNZ) The plans or the form of something.

edges (EJ-ez) The line segments where two faces, including the bases, of a solid figure meet.

faces (FAYS-ez) Any of the polygons that form the sides and bases of a solid figure.

lateral edges (LAT-e-rel EJ-ez) In a solid figure, the line segments formed where the sides of the solid meet each other.

lateral faces (LAT-e-rel FAYS-ez) In a solid figure, the polygons that form only the sides.

net (NET) A two-dimensional pattern for a three-dimensional figure.

polygon (PAH-lee-gon) A plane, closed, many-sided figure.

surface area (SER-fis AYR-ee-uh) In a solid figure, the sum of the areas of each of the faces.

three-dimensional (three-deh-MENCH-nul) Having length, width, and height.

vertex (VER-teks) The shared point where two lines meet.

Index

Web Sites

Due to the changing nature of Internet links, PowerKids Press has developed an online list of Web sites related to the subject of this book. This site is updated regularly. Please use this link to access the list:
www.powerkidslinks.com/psgs/pyramids/